| OUT OF THIS WORLD |

Meet NASA Inventor Joel Sercel and His Team's

Sun-Powered Asteroid Miners

www.worldbook.com

World Book, Inc.
180 North LaSalle Street
Suite 900
Chicago, Illinois 60601
USA

For information about other World Book publications, visit our website at www.worldbook.com or call 1-800-WORLDBK (967-5325).

For information about sales to schools and libraries, call 1-800-975-3250 (United States), or 1-800-837-5365 (Canada).

© 2024 (print and e-book) by World Book, Inc. All rights reserved. No part of this publication may be reproduced, stored in a retrieval system, or transmitted in any form or by any means (electronic, mechanical, photocopying, recording, or otherwise) without written permission from World Book, Inc.

WORLD BOOK and the GLOBE DEVICE are registered trademarks or trademarks of World Book, Inc.

Produced in collaboration with the National Aeronautics and Space Administration (NASA).

Library of Congress Cataloging-in-Publication Data for this volume has been applied for.

Out of This World
ISBN: 978-0-7166-6564-9 (set, hc.)

Sun-Powered Asteroid Miners
ISBN: 978-0-7166-6572-4 (hc.)

Also available as:
ISBN: 978-0-7166-6580-9 (e-book)
ISBN: 978-0-7166-6588-5 (soft cover)

Staff

Editorial

Vice President
Tom Evans

Senior Manager, New Content
Jeff De La Rosa

Writer
William D. Adams

Editor
Emma Flickinger

Curriculum Designer
Caroline Davidson

Proofreader
Nathalie Strassheim

Indexer
Nathaniel Lindstrom

Graphics and Design

Senior Visual
Communications Designer
Melanie Bender

Digital Asset Specialist
Rosalia Bledsoe

Acknowledgments

Cover	© Trans Astronautica Corporation	24-25	© Interfoto/Alamy Images; © Mario Suriani, AP Photo
3	© Trans Astronautica Corporation; © Jurik Peter, Shutterstock	26-27	© Trans Astronautica Corporation
		28-29	© Trans Astronautica Corporation; WORLD BOOK
4-5	ESO	30-31	© Dima Zel, Shutterstock; © Trans Astronautica Corporation
6-7	© Trans Astronautica Corporation	32-33	Joel Sercel
8-9	© Jurik Peter, Shutterstock	34-35	© Trans Astronautica Corporation
10-11	© Mike Ver Sprill, Shutterstock; Joel Sercel	36-37	Joel Sercel/NASA
12-13	© Nazarii Neshcherenskyi, Shutterstock	38-41	© Trans Astronautica Corporation
14-21	© Trans Astronautica Corporation	42-43	© Trans Astronautica Corporation; WORLD BOOK
22-23	© Mopic/Shutterstock	44	Joel Sercel

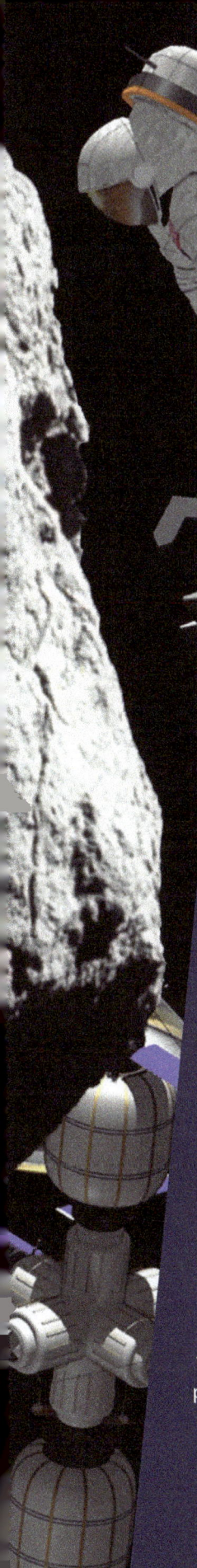

Contents

- **4** Introduction
- **8** The case for mining water
- **10** INVENTOR FEATURE: Overview on Earth
- **12** Near-Earth asteroids
- **14** Sutter: Starting the gold rush to space
- **18** BIG IDEA: Economic sustainability
- **20** Sutter Ultra
- **24** INVENTOR FEATURE: Science fiction inspiration
- **26** Traveling to NEA's
- **28** Concentration over conversion
- **30** On Omnivore's menu
- **32** INVENTOR FEATURE: Self-motivated
- **34** Capturing the asteroid
- **36** Resource harvesting
- **38** Mini Bee experiment
- **40** The impact of mining
- **44** Joel's team
- **45** Glossary
- **46** Review and reflect
- **48** Index

Glossary There is a glossary of terms on page 45. Terms defined in the glossary are in boldface type that **looks like this** on their first appearance on any spread (two facing pages).

Pronunciations (how to say words) are given in parentheses the first time some difficult words appear in the book. They look like this: pronunciation (pruh NUHN see AY shuhn).

Introduction

For much of the world's population, it's a great time to be alive. Across the globe, people have better access to nutritious food, medicine, clean drinking water, and shelter than at any time in human history. Technology enables people to live to a standard that might have been thought impossible only a few decades ago.

There is plenty of room for improvement, however. Many of us would like to see the benefits of modern living spread more thoroughly to every one of Earth's 8 billion people. It is also very human to hope that every member of each future generation enjoys better lives than those before, with all the things they will need to grow, thrive, and be happy.

To spread the best that modern living has to offer without taxing Earth's resources and harming the environment too much, however, we need to be even more clever with our science and technology. We need to invent even better ways to grow and thrive while reducing our impact on Earth, with all its beautiful plants and animals.

For visionaries like the inventor Joel Sercel, one clear way to do that is to head into the heavens—to go to outer space, to take advantage of the virtually unlimited resources the universe has to offer.

> **Asteroids** are so important because they're spread throughout the **solar system** as a fantastic resource for humanity. —Joel

" For the long-term future of humanity, we need to harvest the resources of space, so that people can continue to grow and thrive and that every generation can have a better life than the generation before. " —Joel

Meet Joel Sercel.

" Ever since I was a child, I was inspired by the future depicted by science fiction authors, illustrators, and television writers and visionary scientists. Now, I'm working to make those visions into reality. "

An **asteroid** is a rocky or metallic object smaller than a planet that **orbits** a star. As the Space Age took off in the 1950's, people began imagining mining the resources available within asteroids. Some even formed asteroid-mining companies. But, what seems like a relatively simple concept is deceptively difficult to do economically.

> I started TransAstra with the idea of building the technologies and systems so that humanity can harvest the resources from asteroids. —Joel

Joel and his team at TransAstra have broken down the huge challenge of mining asteroids into four smaller technical challenges: 1) finding candidate asteroids, 2) traveling to and from asteroids, 3) capturing asteroids, and 4) processing asteroids.

> What TransAstra has done through the NIAC program is to develop major inventions in each of those areas. Those inventions will enable humanity to live and work in space, harvesting the resources of the asteroids. —Joel

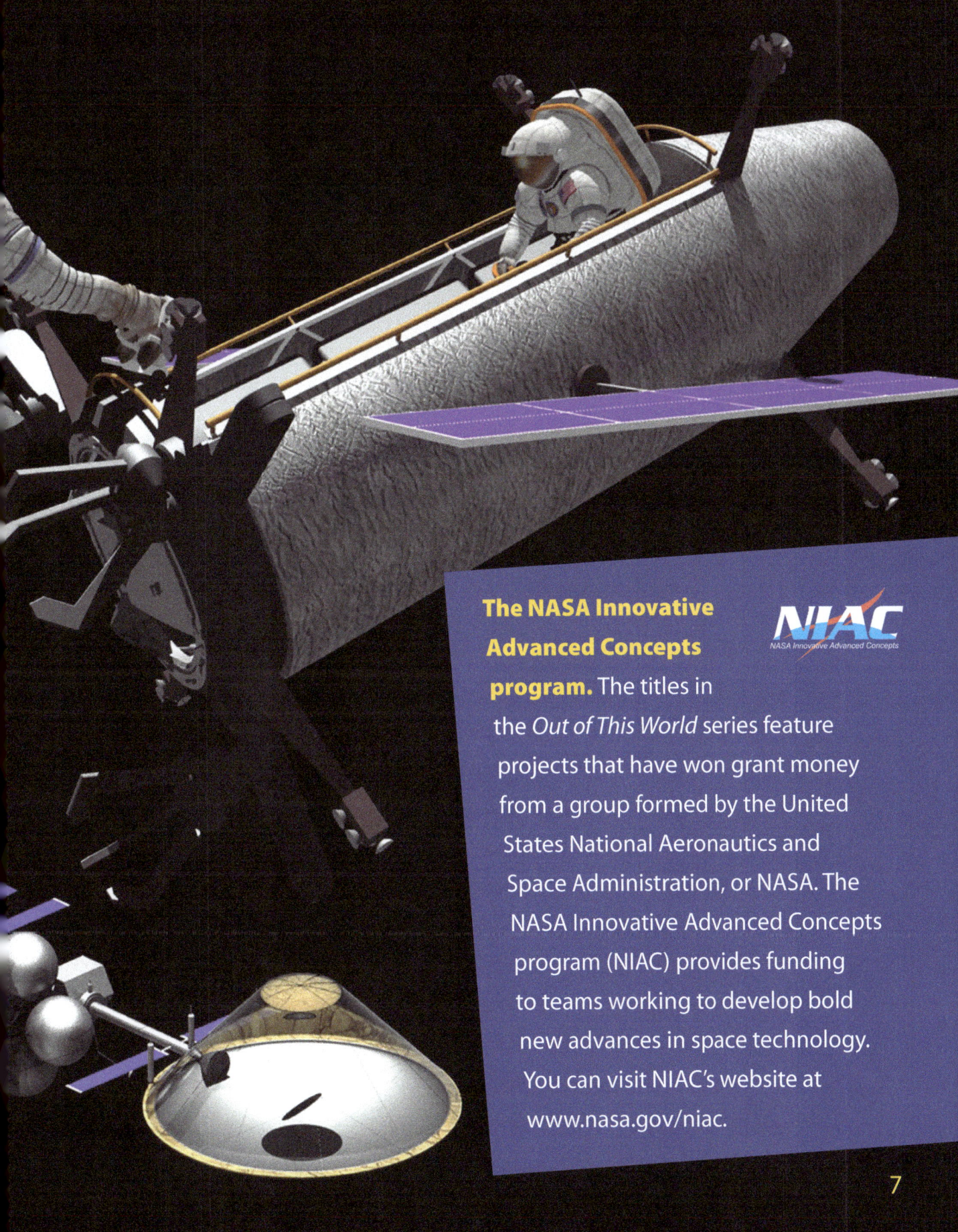

The NASA Innovative Advanced Concepts program. The titles in the *Out of This World* series feature projects that have won grant money from a group formed by the United States National Aeronautics and Space Administration, or NASA. The NASA Innovative Advanced Concepts program (NIAC) provides funding to teams working to develop bold new advances in space technology. You can visit NIAC's website at www.nasa.gov/niac.

The case for mining water

Asteroids contain many minerals that are extremely useful or rare on Earth, including iron, gold, and palladium. We will get those resources from asteroids some day. But TransAstra plans to start by mining asteroids for water. Earth is covered in water. Why go to such trouble to harvest such a common resource?

Although there's a lot of water on Earth, think of how important it is. All living things need water to survive. Most products use water at some

point in their production. And, as important as water is on Earth, it is even more important in space. Of course, future space colonists will need water to drink and to grow crops. But it has other uses, as well.

Each water **molecule** is composed of two hydrogen **atoms** and one oxygen atom. Through the use of electricity, water can be split into oxygen and hydrogen gases. Oxygen can be used for colonists to breathe. Hydrogen can be made into rocket **propellant** (fuel).

People will certainly extract other resources from asteroids. But to do so, we will have to be able to harvest water from them first.

Artist's depiction of plumes of water vapor erupting from an ice-covered body

Inventor feature:
Overview on Earth

Joel grew up in the Arizona desert, where his father was a pilot for the United States Air Force.

❝ I remember seeing his jet plane zooming through the night sky like a red flare. ❞ —Joel

Arizona's high elevation, clear weather, and little **light pollution** make it a perfect place to look at the stars. Growing up in such a fantastic dark-sky environment, along with his father's night-flying, gave Joel a unique perspective on Earth's position in the universe.

Joel's father, U.S. Air Force pilot Major John Sercel

❝ A lot of people think, 'Well, there's Earth and there's space.' Usually, when astronauts go into space, they start to see that Earth is *in* space and that we're all citizens of the universe. That's called the *overview effect*… I came into the world more with that overview effect because I lived in a place where at night you could see space coming right down to the surface. I saw my father flying through the night sky. So, it just seemed obvious to me that that was where we needed to go. ❞ —Joel

Near-Earth asteroids

In our **solar system,** most **asteroids orbit** the sun between the orbits of Mars and Jupiter. This region is known as the **main belt.** These asteroids are too far out of the way for early mining operations. It would take years to fly to them and return.

Asteroids exist all over the solar system, however. Joel and his team are interested in a class of asteroids called **near-Earth asteroids** (NEA's). NEA's have orbits that pass close by the orbit of Earth. Some asteroids even cross Earth's path around the sun. Because NEA's are much closer than the main belt, mining craft can reach them and return more quickly, all while using less fuel.

There are plenty available, too. Astronomers estimate that there are nearly 1 billion NEA's the size of an automobile or larger. TransAstra has also calculated that there are likely thousands of NEA's the size of a house or larger that are easier to reach than the moon.

But, we have only found a few thousand of the estimated 1 billion NEA's.

> **❞ The reason we haven't discovered the rest is because our telescopes aren't good enough. ❞** —Joel

A camera gathers light from the scene being photographed in a process called an *exposure*. In photography, *exposure time* is how long a camera lets light in for a photograph. If you take a short-exposure photograph of space, only the brightest objects would be visible. A longer exposure allows for dimmer objects to appear. But, fast-moving dim objects dart between camera *pixels* (image units) during a long exposure.

Sutter:
Starting the gold rush to space

We cannot mine **asteroids** that we cannot find. So, Joel and his company have developed an advanced telescope system to identify the missing tiny, dark objects. The system is named Sutter.

In the Sutter system, a telescope takes many pictures per minute. If the telescope photographs an unidentified dim patch of pixels, a built-in computer calculates many possible paths that a potential object could take. If more patches show up along one of these calculated paths, then the object is flagged as a potential asteroid.

This technique is called *synthetic tracking*. Joel's team has developed special software that can perform the calculations more quickly and efficiently than ever before. They call the new software *Theia,* after an ancient Greek goddess of light and vision.

One way to conduct synthetic tracking is using a powerful telescope and supercomputer. But with Theia, TransAstra can make use of many identical, low-cost, off-the-shelf telescopes and everyday computer processors. Such equipment can be purchased cheaply in bulk from manufacturers, avoiding a lengthy and expensive development process. This will make it affordable for TransAstra and NASA to find many more asteroids in a short amount of time.

Sutter Ground Network scans the sky at the Winer Observatory in Arizona.

❝ Sutter is named after a mill in northern California where they originally discovered gold. That led to the California Gold Rush. ❞ —Joel

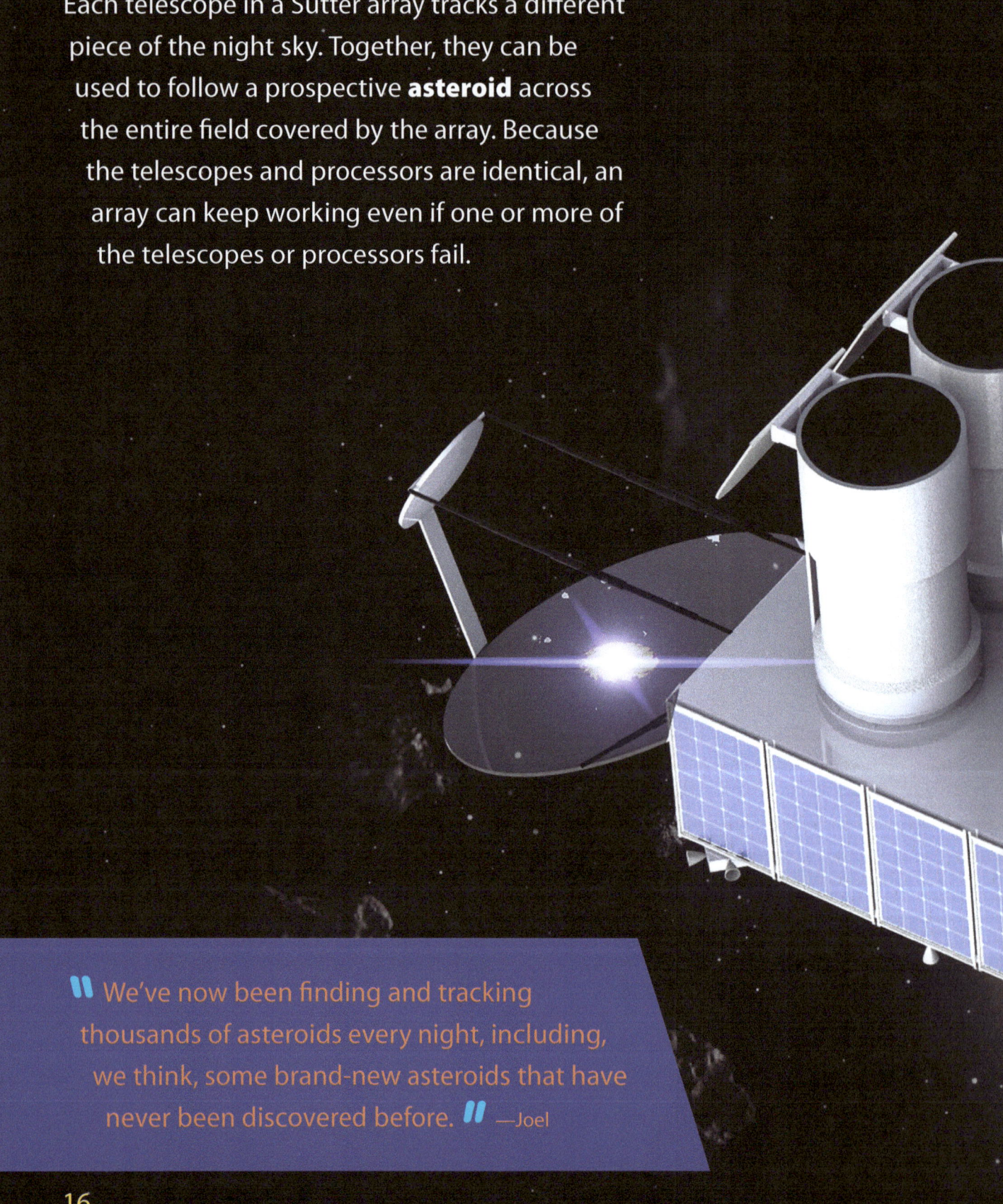

Each telescope in a Sutter array tracks a different piece of the night sky. Together, they can be used to follow a prospective **asteroid** across the entire field covered by the array. Because the telescopes and processors are identical, an array can keep working even if one or more of the telescopes or processors fail.

> ❝ We've now been finding and tracking thousands of asteroids every night, including, we think, some brand-new asteroids that have never been discovered before. ❞ —Joel

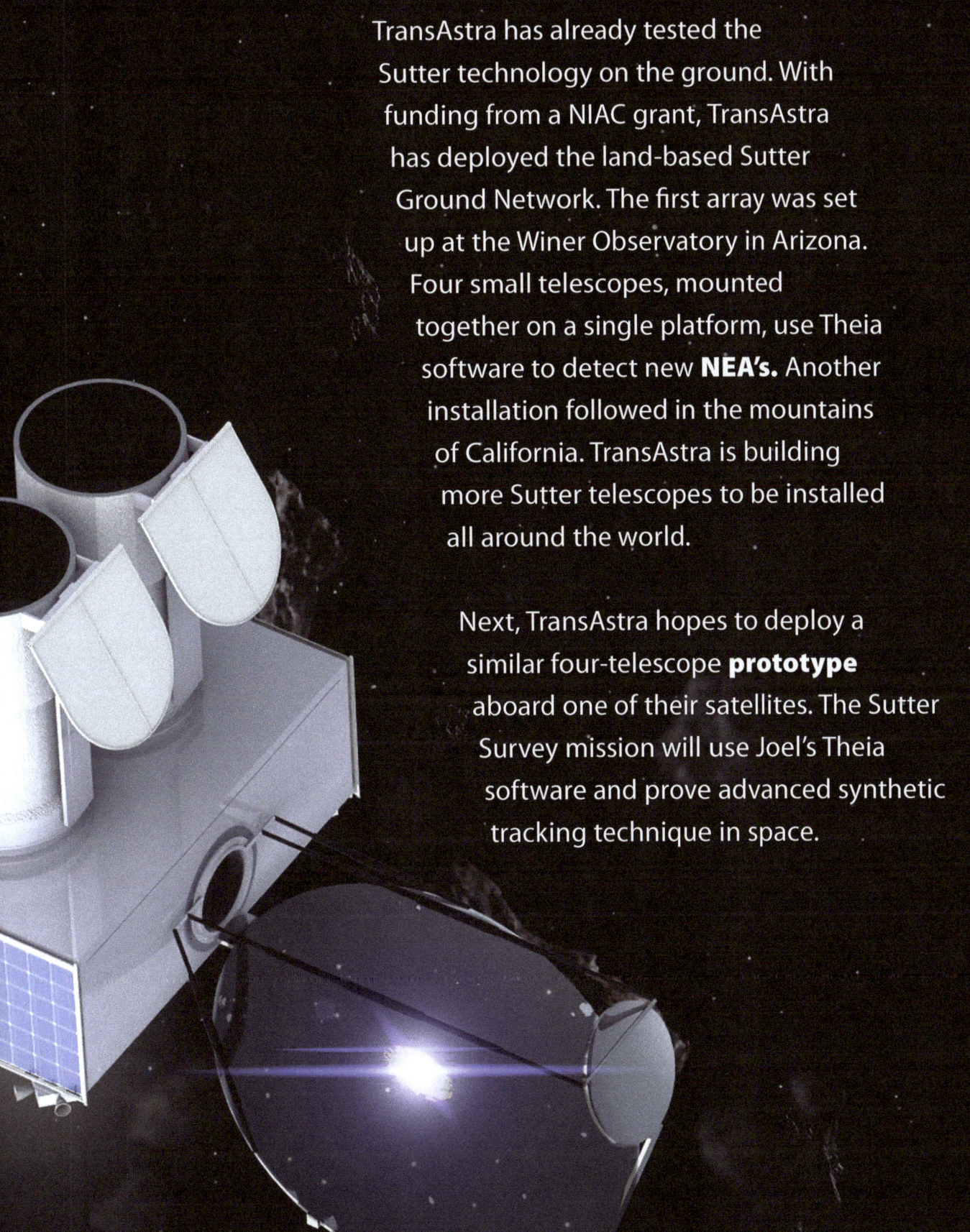

TransAstra has already tested the Sutter technology on the ground. With funding from a NIAC grant, TransAstra has deployed the land-based Sutter Ground Network. The first array was set up at the Winer Observatory in Arizona. Four small telescopes, mounted together on a single platform, use Theia software to detect new **NEA's.** Another installation followed in the mountains of California. TransAstra is building more Sutter telescopes to be installed all around the world.

Next, TransAstra hopes to deploy a similar four-telescope **prototype** aboard one of their satellites. The Sutter Survey mission will use Joel's Theia software and prove advanced synthetic tracking technique in space.

Big idea:
Economic sustainability

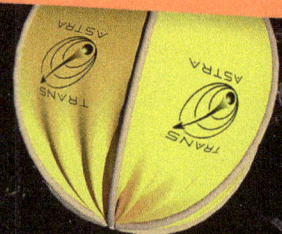

Asteroid mining is a fantastically complex challenge. Not only does a company have to overcome technological hurdles to identify, travel to, capture, and process an asteroid. It must also do so economically. Many asteroid-mining companies have simply run out of money before they could develop their technology.

NIAC funding has helped TransAstra move forward.

❝ It's wonderful that NASA and the NIAC program help do the early-stage development of these technologies. ❞ —Joel

Of course, TransAstra needs more funding than NIAC can provide. But, investors want to see returns in years, not decades. Therefore, TransAstra is developing additional commercial applications in

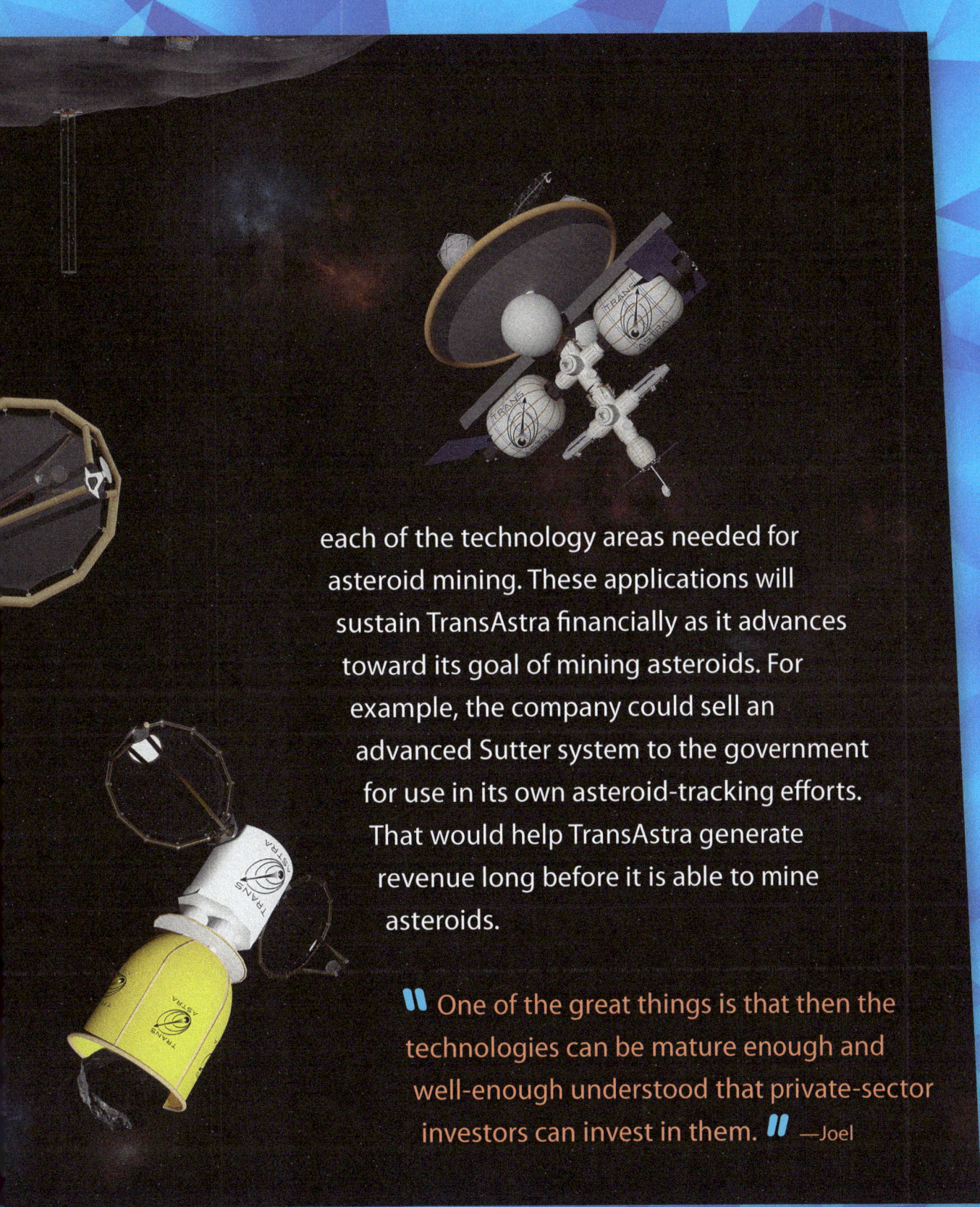

each of the technology areas needed for asteroid mining. These applications will sustain TransAstra financially as it advances toward its goal of mining asteroids. For example, the company could sell an advanced Sutter system to the government for use in its own asteroid-tracking efforts. That would help TransAstra generate revenue long before it is able to mine asteroids.

❚❚ One of the great things is that then the technologies can be mature enough and well-enough understood that private-sector investors can invest in them. ❚❚ —Joel

Sutter Ultra

Sutter Ultra is the final form of TransAstra's Sutter technology. Each Sutter Ultra satellite will carry 109 30-centimeter (12-inch) telescopes. The telescopes' fields of view will be arrayed to form a cross, and the satellite will rotate. Each satellite will be able to cover an enormous portion of the sky.

Sutter Ultra could have a profound effect on astronomy. Joel's team estimates that an array of three Sutter Ultra satellites will discover 300 times as many **asteroids** in the first year of operation than have been discovered in the history of astronomy.

❚❚ Our Sutter space telescopes can be used not just to find and track asteroids, but also to find and track *orbital debris* (broken-up pieces of spacecraft) or spacecraft that have stopped working correctly or that have been lost. That process is called *space domain awareness* (SDA). There's an important commercial market for SDA to help business and government satellites understand where other satellites are to avoid collisions. It can also help us keep space tidy by finding debris that can be cleaned up. ❚❚ —Joel

Sutter Ultra could be profitable for TransAstra in the developing field of space domain awareness.

Tracking space debris is even more critical than tracking other satellites. As more and more launches go into space, they release more and more bits of debris. These debris pieces are usually tiny. But objects in space travel at high speeds—many thousands of miles or kilometers per hour. At such speeds, a collision with even a small piece of debris could destroy a satellite, creating even more debris.

Some people worry that a chain reaction of such collisions could render entire regions of **orbit** unusable. Such a proposed event is called Kessler Syndrome, for the scientist who first proposed it, Donald Kessler.

Three Sutter Ultra satellites positioned at roughly the orbit of the moon could track just about every object in orbit around Earth.

Sutter Ultra could serve as a last line of warning against **asteroids** that could hit Earth. If a dangerous asteroid was approaching Earth, the array could detect it and give people vital hours to prepare.

Inventor feature:
Science fiction inspiration

Growing up, Joel was inspired by science fiction.

❝ I got to stay up way past my bedtime to watch 'Star Trek.' The idea that there was this unlimited future for humanity out in space just always seemed obvious to me. ❞ —Joel

❝ The great science-fiction writers of the 20th century, including Robert Heinlein, Arthur C. Clarke, and Isaac Asimov… I read all their books as a child— every one of them. Those had profound effects on me. I read science fiction voraciously. ❞ —Joel

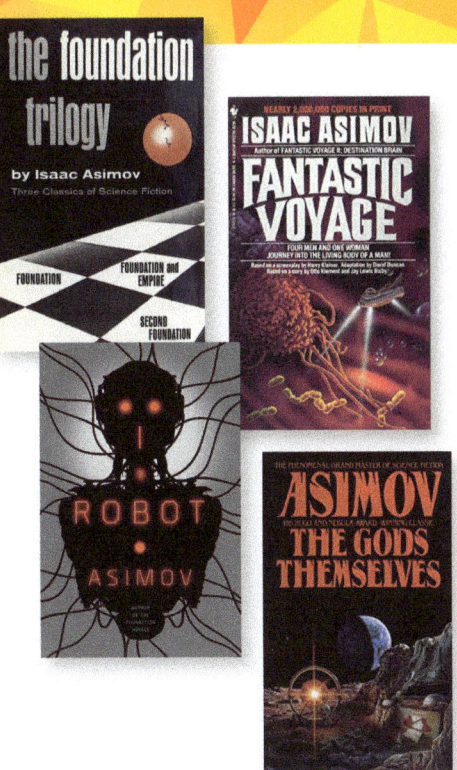

Isaac Asimov (1920-1992) was an American author. He wrote about 400 books for young people and adults, mostly nonfiction emphasizing science and technology. However, he became best known for his science fiction. Many of Asimov's short stories and novels feature robots as characters. Several were collected in *I, Robot* (1950). His popular Foundation series of science-fiction novels includes *Foundation* (1951), *Foundation and Empire* (1952), *Second Foundation* (1953), *Foundation's Edge* (1982), *Foundation and Earth* (1986), *Prelude to Foundation* (1988), and *Forward the Foundation* (published in 1993, after his death). He also wrote *Fantastic Voyage* (1966) and *The Gods Themselves* (1972).

Traveling to NEA's

The second technical challenge of **asteroid** mining is traveling to and from the target asteroid.

Engineers often choose *ion propulsion* for a mission that involves a long cruise across the **solar system.** In ion propulsion, a fuel is sent through an electrified grid. This heats the fuel and *ionizes* it, turning it into electrically charged

In this image captured from a video, the Omnivore propulsion system is tested at TransAstra's laboratory.

particles. The charged particles then exit through the nozzle at a high velocity, pushing the craft in the opposite direction.

Ion propulsion is more efficient than conventional chemical rocket propulsion. But, it requires massive solar panels to collect the energy necessary to propel the craft. Such solar panels increase the launch **mass** of the spacecraft, reducing the profitability of a commercial mission.

Joel has invented a propulsion system called Omnivore. Omnivore is a type of **solar thermal propulsion.**

> ❝ The Omnivore propulsion system is a new type of solar thermal rocket. It uses lightweight solar concentrators that are very affordable and efficient to gather sunlight onto a small spot in a rocket engine. When you concentrate sunlight onto a small spot, you can create very, very high temperatures. ❞ —Joel

Propellant passed through this heated spot rapidly expands. It shoots out the back of the rocket engine, producing **thrust.**

> ❝ It heats up the **thruster** to a very high temperature, and then we can put virtually any fluid into the thruster to use as propellant. ❞ —Joel

Concentration
over conversion

Ion **propulsion** and **solar thermal propulsion** appear similar at first glance. Both heat, rather than combust, a **propellant** to create **thrust.** Why go to the effort of developing a different system?

The answer lies in the fundamental laws of physics. Energy transfers are not perfectly efficient. Any time energy changes from one form to another, some of it is converted to **friction,** waste heat, or other unuseful forms.

Solar panels on spacecraft are about 34 percent efficient at converting sunlight into electricity, so two-thirds of the energy is wasted. But Omnivore concentrates, rather than converts, sunlight. Because there is less waste, the concentrators needed are less massive than the solar panels needed for an ion rocket of the same size.

TransAstra is utilizing another novel technology to make their concentrators even lighter. The concentrators will be inflatable! After launch, a small amount of gas, called *inflatant,* will inflate the concentrators to the necessary shape.

Inflatables are finding all kinds of uses in astronomy and space exploration. You can read more about NIAC fellow Chris Walker's plans for such technology in *Out of This World: Inflatable Stargazers!*

29

On Omnivore's menu

The word *omnivore* means something that eats all kinds of food—think of a bear, or a person. The Omnivore **thruster** can use a variety of **propellants,** including water. This is a big difference from other **propulsion** systems, which are fine-tuned to work with a specific propellant.

The Omnivore thruster can run on water. This means that a TransAstra mining craft can mine its own return fuel! This ability makes it more economical and extends its range. The craft only needs to carry enough fuel to reach a target **asteroid.** Once there, it can mine the asteroid and use some of the harvested water to return to its base.

In keeping with TransAstra's principle of ensuring profitability at every point in the technology's development, the company is developing an Omnivore-powered spacecraft for use as a space tug. Omnivore's strong, efficient **thrust** makes it useful anywhere within the **main belt.**

❚❚ The Omnivore propulsion system has commercial applications in the very near term to be used as an upper stage for small rockets and to carry **payloads** in space from one spot to another, both for the government and for private businesses. ❚❚ —Joel

Inventor feature:
Self-motivated

Joel was bored at school. He was not interested in many of the classroom subjects and had difficulty grasping them the way they were being taught.

Joel on his paper route

> I happened to come from a family where most people are naturally intelligent, but I also had learning disabilities. So I had a hard time learning in schools but an easier time teaching myself outside the school environment. —Joel

Joel would skip class and go to the library and read books about science and technology. Or, he would pretend to be sick to stay home and work on inventions.

❝ If you're waiting for someone else to teach you stuff, the best you're going to learn is what everyone else knows. ❞ —Joel

As Joel got interested in **asteroid** mining in high school, he read about the Orion Project. This project would have used nuclear detonations to propel spacecraft at extremely high speeds. Joel wrote to the person who came up with the idea, the physicist Freeman Dyson, asking if the Orion Project technology could be used for asteroid mining. Dyson wrote back, beginning a correspondence and friendship that lasted until Dyson's death in 2020.

Freeman Dyson

Joel's interest in asteroid mining and rocket **propulsion** sparked his pursuit of higher education. He earned a bachelor's degree from the University of Arizona and a Ph.D. degree from the California Institute of Technology.

Capturing
the asteroid

The third technical challenge of **asteroid** mining is capturing the asteroid.

An object's **gravitational force** is related to its size. Think about how the Apollo program astronauts could jump on the moon—despite being in their heavy space suits. Smaller planetary bodies have even less gravity. Asteroids the size TransAstra is targeting cannot be landed upon in the conventional sense. The asteroid's gravity is too weak to hold a craft to it. Furthermore, because of the weak gravitational pull, materials can easily escape into space during the mining process. This could result in lost revenue and produce debris clouds hazardous to the asteroid miner.

> So, if you're going to do **engineering** work with the material, the first thing you need to do is capture it in a bag. —Joel

Many thinkers have proposed capturing the asteroid with a large, unfolding membrane (or "bag"). TransAstra is working on this technology, too. TransAstra's breakthrough is devising other uses for its capture technology.

Sometimes, satellites malfunction or go offline, posing a hazard to other satellites and spacecraft. A tug equipped with a capture bag

In this rendering, a TransAstra satellite uses a capture bag to collect an asteroid for processing.

could grab such a satellite and move it for disposal—even if there's no control of it or it lacks any kind of docking port. By pairing bag-capture capabilities with Sutter telescope arrays, the company will be able to locate, capture, and deorbit broken satellites and pieces of debris, making space safer for satellites and spacecraft.

> We can use our smaller capture bags—before we have the really big ones for capturing asteroids—to go clean up orbital debris. There are both commercial and government opportunities to do that in the near future. —Joel

Resource harvesting

The final, most important technical challenge of **asteroid** mining is harvesting the materials.

> ❝ To extract the water from the asteroid, all you have to do is heat it up. ❞ —Joel

Other groups have thought to extract the materials by heating the asteroid to high temperatures within the capture bag. But TransAstra has determined that it would take too much energy and time to heat an entire asteroid all at once.

Instead, TransAstra has developed a technique called *optical mining.* By moving small mirrors within the craft, the same solar collectors used for the Omnivore **thruster** concentrate sunlight to a tiny point on the asteroid. The concentrated sunlight heats the point to extremely high temperatures. Such heated volatiles as water ice vaporize and explode out of the rocky asteroid, knocking chunks off its surface. The volatiles can then be collected within the spacecraft and stored in tanks.

It is hard to believe that you can drill through solid rock using just the power of sunlight—without even touching the rock! But Joel and his team have already done it in the lab. They have drilled through dozens of materials, including simulated asteroid material and even material from actual meteorites.

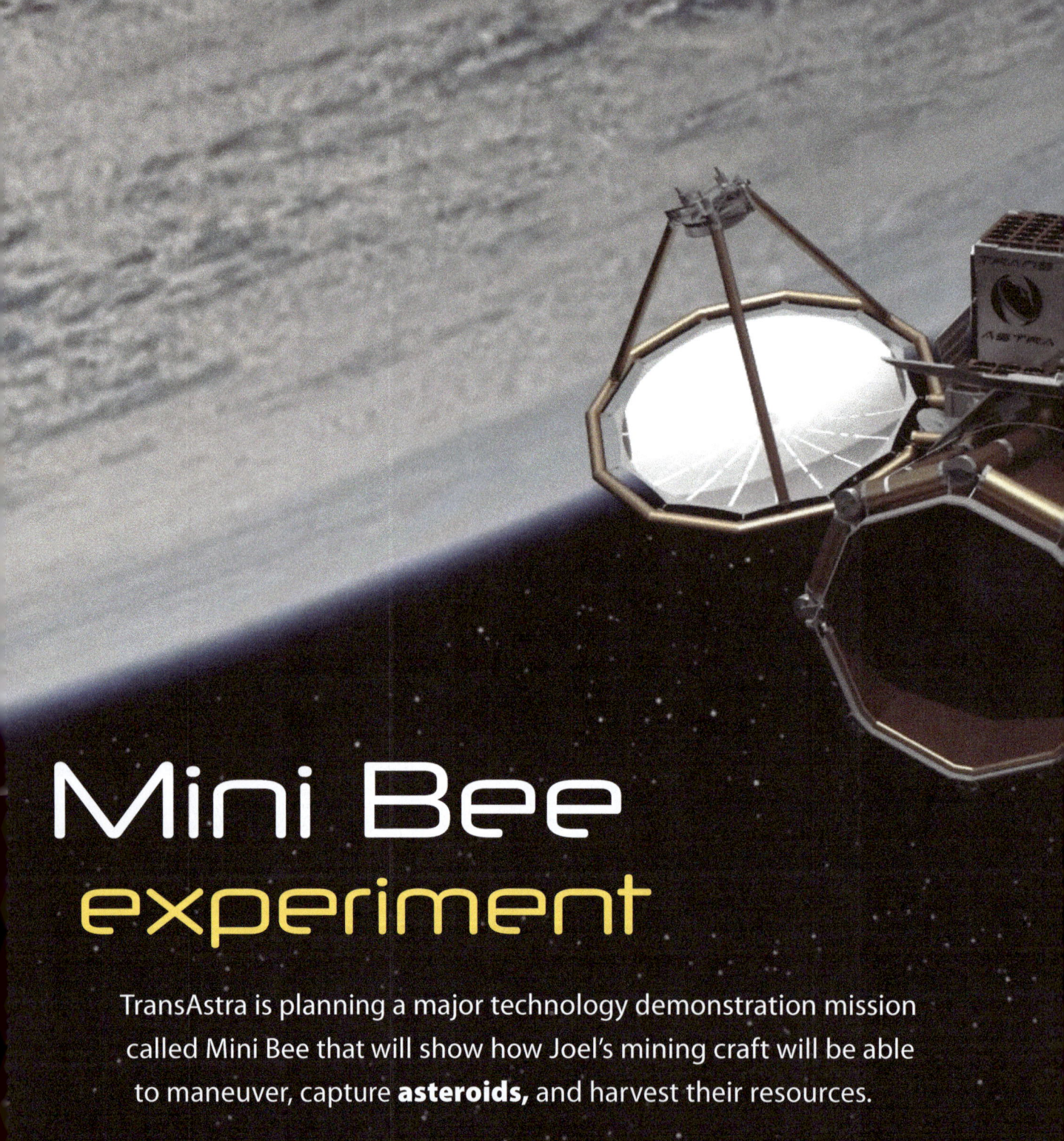

Mini Bee
experiment

TransAstra is planning a major technology demonstration mission called Mini Bee that will show how Joel's mining craft will be able to maneuver, capture **asteroids,** and harvest their resources.

The Mini Bee will be boosted into Earth **orbit** along with a *synthetic* (lab-created) asteroid. Using Omnivore **thrusters,** the Mini Bee will catch up to the asteroid and maneuver to match its spin. Then, it

Artist's rendering of the Mini Bee satellite in orbit around Earth

will deploy its capture bag and collect the asteroid. After it collects the asteroid, it will demonstrate the optical mining technology, capturing volatiles locked in the sample.

After this Mini Bee trial, TransAstra will develop a larger satellite, called the Worker Bee, which will serve as its primary mining craft. It will travel to **NEA's** identified by Sutter Ultra to harvest their water.

The impact of mining

After the Worker Bee has mined through the whole **asteroid,** it will return to Earth **orbit** with tanks full of water. This water will be the foundation of a new space economy, with a number of key uses.

Life support. Astronauts and colonists in space can drink water and use it to grow food crops. Oxygen derived from water can be added to spacecraft or space **habitats** for people to breathe.

Water as fuel. Currently, a rocket must haul all its fuel for the entire mission up with it from Earth. And, the more fuel a rocket needs to carry, the more fuel it takes to lift all that extra fuel. **Engineers** call this the "tyranny of the rocket equation."

Water mined from asteroids will be electrolyzed into hydrogen and oxygen gas at an orbiting factory. The Omnivore **propulsion** system runs most efficiently on pure hydrogen fuel. So, the Worker Bee could be refueled with refined hydrogen in space and sent out to its next mining target. Hydrogen and oxygen can also be used as fuel for chemical rockets.

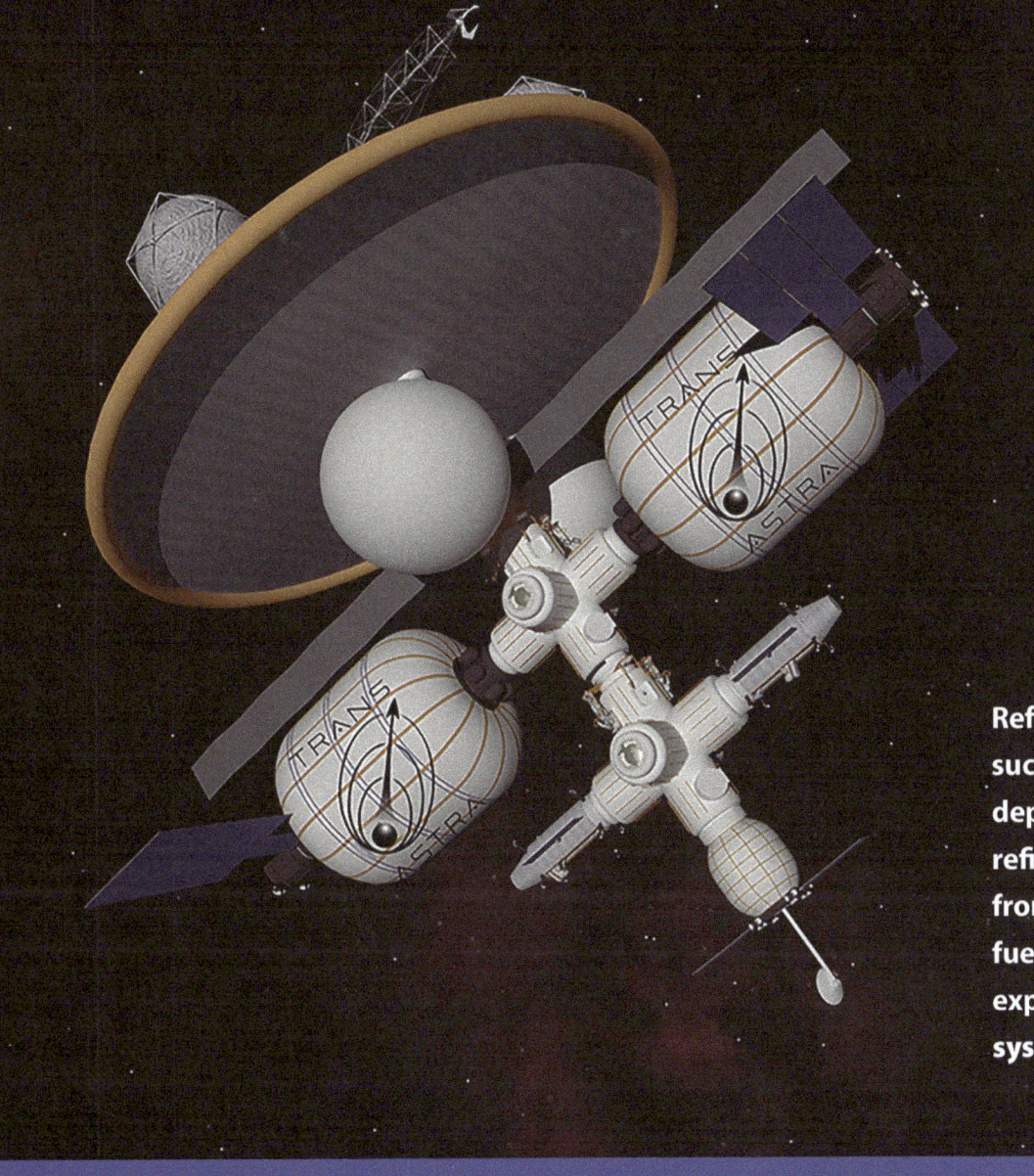

Refueling stations, such as the one depicted above, could refine water mined from asteroids into fuel for spacecraft exploring the solar system.

A refueling station in Earth orbit would enable mission planners to break free from the tyranny of the rocket equation. Each mission would only need to be loaded with the fuel required to get it to the refueling depot. Any mission, from satellite space **probes** to crewed missions to Mars, could be larger, faster, and cheaper.

❚❚ Fifty, or 100, or 200 years from now, we could see millions, billions, or tens of billions of people living in space in worlds made out of **asteroids.** Some of them could live in **orbit** around Earth or the moon. Some of them could live in deep space. And they could live in places spread from the orbit of Venus all the way to the orbit of Jupiter. ❚❚ —Joel

In-space manufacturing. Asteroids are full of other valuable materials. These materials might be used to make products for space missions or colonies on the moon, Mars, or in orbit.

Radiation shielding. Earth's **atmosphere** protects us from much harmful **radiation** from the sun. In space, this shielding is absent. But rock and liquid water also have good radiation shielding properties.

Architect Anthony Longman has won a NIAC grant for his proposal to build a space **habitat** that can expand to hold thousands of people. His proposal calls for layering bladders of water and leftover mining "slag"—both products of asteroid mining—around the outside of the habitat to protect the inhabitants from radiation. You can check out Anthony's idea in *Out of This World: Expandable Space Habitat!*

Joel's team

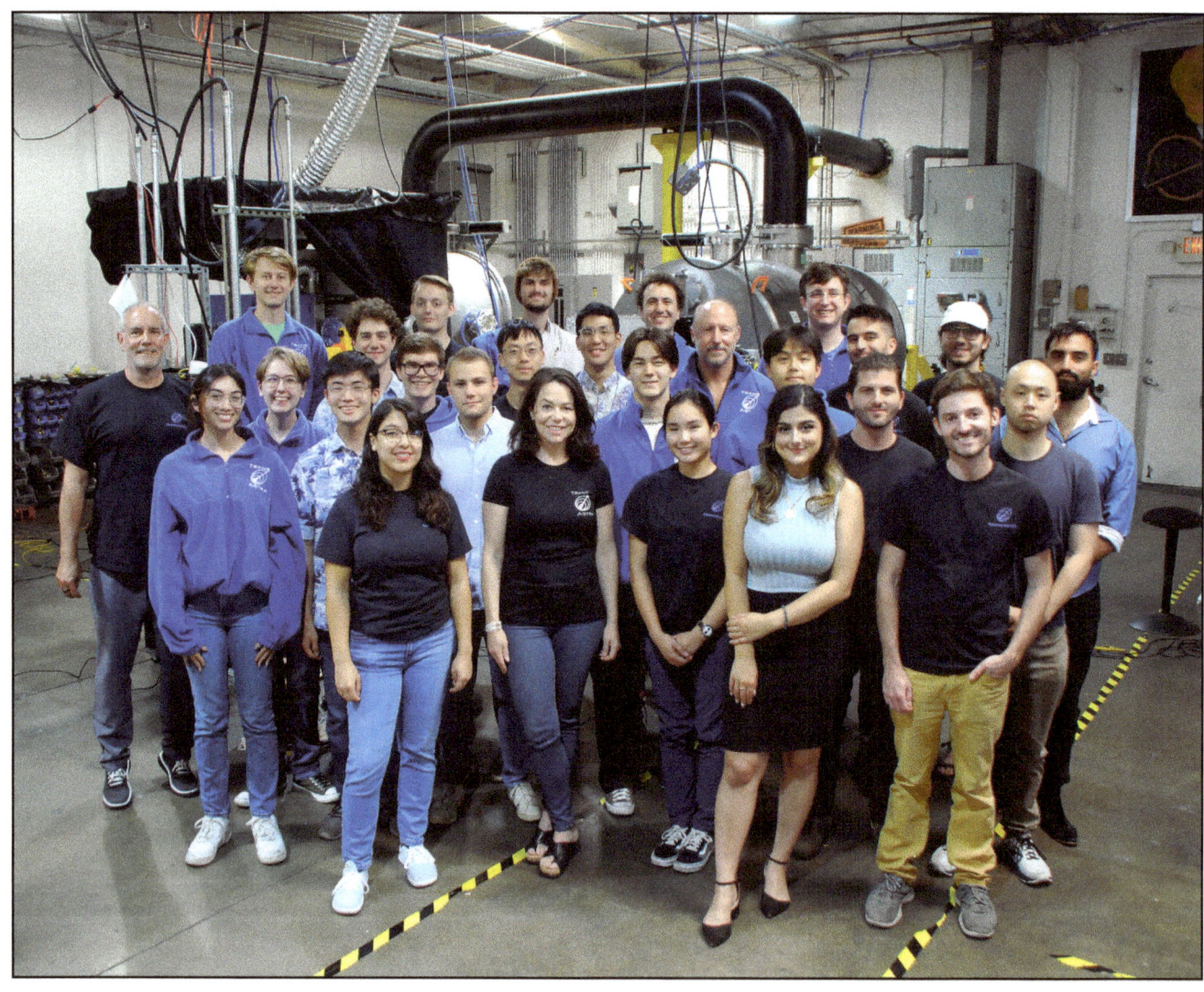

Joel with the staff of TransAstra

Glossary

asteroid a rocky or metallic body smaller than a planet that orbits the sun.

atmosphere the mass of gases that surrounds a planet.

atom one of the most basic units of matter, consisting of a *nucleus* (core) of particles called *protons* and *neutrons* with tiny particles called *electrons* moving around the nucleus.

engineer a person who uses scientific principles to design structures, such as bridges and skyscrapers, machines, and all sorts of products.

friction the resistance of a body in motion to the air, water, or other medium through which it travels or to the surface on which it travels.

gravitational pull also called gravitation or the force of gravity, the force of attraction that acts between all objects because of their mass. Because of gravitation, an object that is near Earth falls toward the surface of the planet. We experience this force on our bodies as our weight.

habitat living space.

light pollution the effects of artificial lighting that harm the environment or interfere with people's view of the night sky.

main belt a band of asteroids circling the sun between the orbits of Mars and Jupiter.

mass the amount of matter something contains.

molecule two or more atoms bonded together.

near-Earth asteroid (NEA) an asteroid that has an orbit that passes close by the orbit of Earth.

orbit a looping path around an object in space; the condition of circling a massive object in space under the influence of the object's gravity.

payload the useful load carried by a vehicle.

probe a rocket, satellite, or other uncrewed spacecraft carrying scientific instruments, to record or report back information about space.

propellant solid or liquid fuel that is turned into gas and put under pressure to push a spacecraft forward.

propulsion pushing something, such as a spacecraft.

prototype a functional experimental model of an invention.

radiation energy given off in the form of waves or tiny particles of matter.

solar system the sun and everything that travels around it, including Earth and all the other planets and their moons.

solar thermal propulsion a type of propulsion system that uses sunlight to directly heat propellant.

thrust moving force; a push with a force.

thruster a rocket or other device used to help steer and control the motion of a spacecraft.

Review and reflect

Now that you've finished reading about Joel Sercel, use these pages to think about his experiences and TransAstra in new ways. As you work, reflect on the importance of creative problem solving, curiosity, and open-mindedness in life.

Complex problems and creative solutions

 Why are scientists and entrepreneurs interested in mining asteroids?

 What are some of the problems associated with mining asteroids?

 How does Joel Sercel hope to overcome these challenges with his company TransAstra? What makes these solutions so innovative?

Visit www.worldbook.com/resources to download sample answers, blank graphic organizers, and a rubric to evaluate writing.

Inspiration can come from anywhere!

Use a graphic organizer like the one below to map out your ideas. What ideas or experiences led to Joel's innovative solutions?

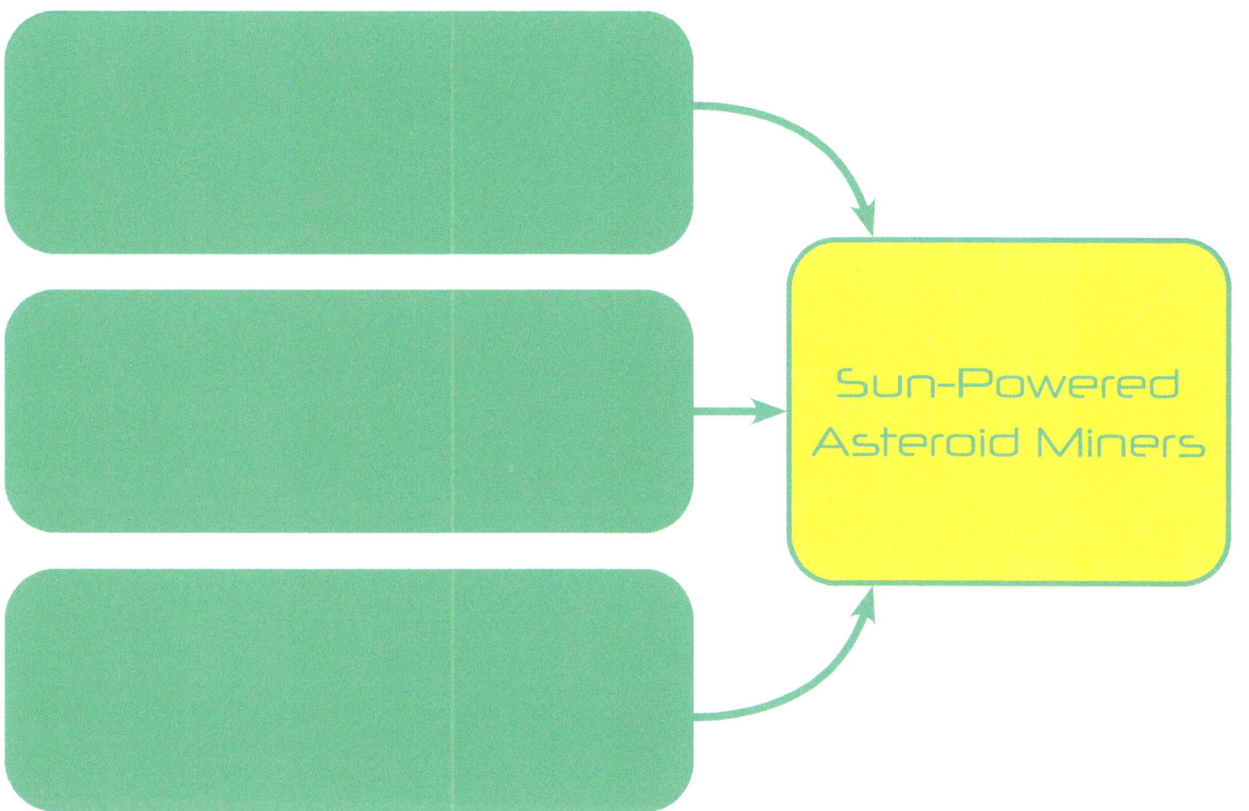

Write about it!

Think about Joel's quote: "If you're waiting for someone else to teach you stuff, the best you're going to learn is what everyone else knows."

Do you agree or disagree with Joel? Can you think of a time when you took it upon yourself to learn or discover something new?

Index

A

Arizona, 10, 15, 17
Asimov, Isaac, 25
asteroid, 6, 8, 12, 16, 20, 22, 26, 30, 34, 36, 42
asteroid mining, 6, 18, 33, 34, 36, 43
astronomy, 20

C

commercial, 18, 21, 26, 31, 35
concentrator, 26, 29

D

Dyson, Freeman, 33

G

gravitational force, 34

H

hydrogen, 9, 40

I

inflatables, 29
ion propulsion, 26, 28

K

Kessler, Donald, 22
Kessler Syndrome, 22

L

Longman, Anthony, 43

M

main belt, 12, 30
manufacturing, 42
Mini Bee, 38

N

NASA Innovative Advanced Concepts program (NIAC), 6, 7, 17, 18, 29, 43
National Aeronautics and Space Administration (NASA), 7, 14
near-Earth asteroids (NEA's), 12, 17

O

Omnivore, 26-27, 29, 30, 31, 36, 38, 40
optical mining, 36
orbit, 12, 22, 38, 40, 41, 42
orbital debris, 21, 34, 35. *See also* space debris
Orion Project, 33
overview effect, 11
oxygen, 9, 40

P

propellant, 9, 27, 28, 30
propulsion system, 30. *See also* ion propulsion; Omnivore

R

radiation, 42

S

satellite, 22, 34, 38-39
Sercel, Joel, 4, 5, 6; education, 32-33; inspiration, 24-25. *See also* TransAstra
science fiction, 24-25
solar panels, 26, 29
solar system, 12, 26
solar thermal propulsion, 26, 28
space debris, 22
space domain awareness, 21, 22
space tug, 30, 34
Sutter (telescope system), 14, 17, 19, 34; Ultra, 20, 22
synthetic tracking, 14, 17

T

telescope, 13, 14, 16-17
Theia (software), 14, 17
thrust, 27, 28
TransAstra, 6, 8, 14, 18-19, 22, 29, 30, 34, 36
"tyranny of the rocket equation", 40-41

W

Walker, Chris, 29
water, 8-9, 30, 36, 40-41, 42, 43
Winer Observatory, 15, 17
Worker Bee, 39, 40

www.ingramcontent.com/pod-product-compliance
Lightning Source LLC
Chambersburg PA
CBHW060934170426
43194CB00023B/2957